Stencilled
Ornament & Illustration

STENCILLED
ORNAMENT & ILLUSTRATION

A Demonstration of
WILLIAM ADDISON DWIGGINS'
Method of Book Decoration
and Other Uses of the Stencil

together with
A Note by the Artist

Compiled and Arranged by
DOROTHY ABBE

Afterword by
BRUCE KENNETT

—

Princeton Architectural Press
New York

for William
"Black and White-Smith"

PREFACE

To imagine the beauty of the delicate stencils which Dwiggins, with incredible skill, fashioned out of thin, transparent materials, would be difficult for anyone who has not seen these plates. The surety of the knife's incision, the bevelled edges of the intricate designs, the originality of their use—these were reasons enough to wish to gather them into a book, although, necessarily, in their ultimate printed form.

But such a compilation interested Dwiggins for quite a different reason. A volume filled with a variety of unrelated lines and curves (building blocks for ornaments, or "elements," as he called them) together with some of the designs formed from these elements, might be of particular interest to a student of art and design.

Thus it was that in the early fifties we decided upon this subject for a book to be produced under the imprint of Püterschein-Hingham, our private press. But his failing health interfered, so that he was not able to do more than write a rough description of the process and make a few sketches. And so our plans were set aside, but not forgotten. Now, some twenty-five years later, the book has taken shape.

Of the various ways in which Dwiggins used the stencil, this book deals chiefly with the most important category: the typographical designs such as vignettes, borders, repeat patterns, etc., made for the most part by combining several small elements; in some instances a single stencil produced the complete design. The elements, reproduced actual size and presented according to Dwiggins' own classifications, are interspersed with examples of ornaments that he made with them. Except for the designs for book covers, and a border for a poster, the ornaments are shown in the sizes in which they appeared originally. Since many of the celluloid stencils have long ago disintegrated, all the components of a specific design cannot always be found among the elements shown. Nevertheless it is interesting to take a certain unit and trace the varied and ingenious ways in which he utilized one simple device.

That these ornaments harmonize so well with type may be attributed in part to the crisp line produced by the knife-cut stencil. However, due to the conditions under which the material for this book was collected, it is inevitable that some reproductions do not exhibit the expected sharpness of line.

Most of the elements shown have been assembled from more than 250 celluloid and acetate stencils, at times there being several elements on one piece of film. These elements were organized as to design, then arranged in an orderly manner by placing them inside ruled panels which are usually page size; a few are smaller or irregular in shape. Thus the elements are separated from the designs made from them. However, the larger units are not so differentiated.

The stencil ornaments here reproduced are but a small selection from those created in that medium. They are characteristic of Dwiggins' style, but not all such ornaments found in his work were made with stencils; some were pen-and-ink, others a combination of stencil and pen.

viii

In widely separated parts of the world the stencil has long been used for both utilitarian and decorative purposes. Two uses are related to the book arts: some choir books of the eighteenth century had text, music, and decorations produced entirely by stencils; at a later date pochoir (the French word for "stencil," denoting the process) was carried to a high degree of perfection in the reproduction of paintings and prints, as well as for the illustration of books.

Thus Dwiggins was utilizing long established procedures in employing stencils for textiles, lettering, and illustration. But his application of this medium to the creation of typographical ornaments was unique, both in the way in which he devised them as well as in the individuality of their style. And that these designs are so harmonious with type—at their sparkling best when printed in black—must lie in the fact that they are the work of a master of calligraphy and type design; or, in his own, more simple words, of a "Black and White-Smith."

Finally, grateful acknowledgement is made to Margaret Evans, Jean Whitnack, Doris Hauman, and Natalie Norris for their encouragement and helpful criticism.

ix

CONTENTS

PHOTOGRAPHS

Stencilled
Ornament & Illustration

A NOTE BY THE ARTIST
CONCERNING HIS USE OF STENCILS
FOR BOOK ORNAMENTATION

THIS experiment with book ornament via stencils had three origins. The first was an attempt along lines that might be called a "rubber stamp prelude." I cut sprigs, leaves, and flowers on several blocks of cherry wood "on the plank." The elements were derived, vaguely, from the backgrounds of Persian book illuminations and from details of blue Canton willow-pattern. With these and a folded-cotton stamp-pad fed with black Higgins drawing ink, I undertook to stamp assembled designs to be reproduced in line by photoengraving. However, this method proved to be too difficult.

The second origin: It occurred to me that stencils cut in thin celluloid would be much easier to make and to place in position because the material was transparent. The edges of the prints* would be sharper. An attempt with acetate stencil sheets instead of celluloid was unlucky. Humidity changes caused the acetate to shrink and swell and so threw multicolor projects out of register. Most of the celluloid stencils kept their shapes, but a number, put away between sheets of paper, underwent some kind of chemical reaction with impurities in the paper that turned both celluloid and paper into brittle scraps. Possibly the use of vinyl would cure both troubles.

The third inspiration occurred underground, in the Park Street

3

---

*Actually, one *stencils* a *stencil* and the product is a *stencil*. For clarity, *print* and synonyms of it are used as needed. —D.A.

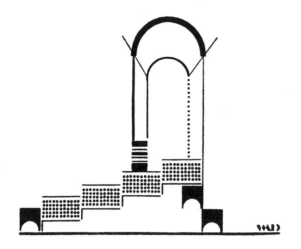

station of the Boston subway. Shadows cast by the concrete beams and arches suggested a battery of elements made up of geometrical shapes and motions—circles, straight lines, triangles, conic section curves. These, cut as stencils, produced what Carl Rollins called my "concrete mixer" style. These elements turned out designs that were novel and lively.

So there were two lines of attack upon ornament via stencils: (1) natural curves, stems, leaves and flowers; (2) geometrical shapes—sharp, hard patterns—sometimes following growth developments, and sometimes abstract geometrical traceries, for their own sweet sake.

It has been interesting to see how the elements could be used to construct pictorial subjects: landscapes, dramatic incidents with figures, etc. The question naturally arises: Why use stencilled scraps for this purpose? Particularly with the geometrical lot, the excuse for the indirect attack is the fact that the assembled elements provide extra-sharp and emphatic statement—a kind of conventionalization arrived at unconsciously, that somehow seems in harmony with type letters. The first making of pictures via stencil was in the Overbrook Press edition of *One More Spring*, 1935. For these semipictorial projects I usually made a rough pencil or watercolor sketch of the action or view, and then applied the stencil interpretation on top of the sketch.

4

For vignettes and other pieces of an ornamental nature I have done without sketch-
ing, and allowed the elements to bud and
grow according to their obvious intentions
and relations. The first use of the stencil
method for making ornament per se occurred in *Paraphs*, Knopf, 1928.

The specimens shown in this book fall into several classes:

What is left * *of the naturalistic celluloid elements*
Geometrical elements and designs
Border elements, both naturalistic and geometrical
"African" decorative elements
Designs to be repeated to make "over-all" patterns
Elements (lines, dots) to be repeated to make textures
Devices without any ornamental aims in themselves,
 cut as aids toward constructing other designs—letter-
 ing, for example, music signs, and signs for maps
Designs for special jobs
Illustration
Examples of elements combined into final form

5

---

* *"What is left"*—apply this throughout. Some varieties
of celluloid stood up against the assaults of chemicals
and time; others crumbled into bits.

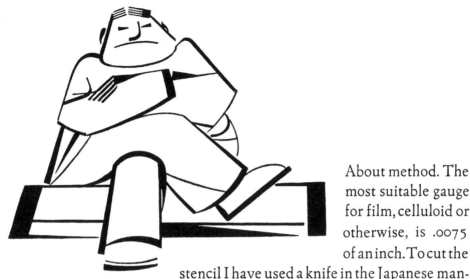

About method. The most suitable gauge for film, celluloid or otherwise, is .0075 of an inch. To cut the stencil I have used a knife in the Japanese manner, pushing it ahead with a finger, instead of dr⁻wing it towards me. The knives were made from hack saw blades by careful chipping, grinding, and whetting—the metal blade laid into two pieces of a split wooden handle and bound about with waxed cord. All of this is very much Japanese.

The cutting is done into bits of celluloid taped in place over the pen drawings of the elements. After the ties have been located, a light cutting is made, not all the way through the film, and if necessary, French chalk rubbed into the scratches. Then over black paper to let you see where to go, a final cut is made through the film.

The stencil brushes are the conventional kind, bought at artists' supply stores, say ⅜" barrel with stiff bristles ½" long.

I brush out Higgins drawing ink on a china slab and charge the stencil brush from this. To charge the brush properly is the critical point in the operation. Too much ink will run out under the stencil plate and destroy the design; too little will call for a dangerous amount of scrubbing. The right amount, found out only by trial and error, will cut the edges sharply. After the stencilling is done, if the design requires it, the ties are filled in with a pen.

WILLIAM ADDISON DWIGGINS

6

SUPPLEMENTARY TECHNICAL DETAILS

Not long after Dwiggins wrote the preceding note, his health became further impaired, thus preventing him from expanding the text or completing the diagrams. The additional details included here are derived from his notes as well as from personal observation and experience.

Dwiggins' stencilled ornaments and illustrations were produced by either of two quite different METHODS. One required that a sketch be made in pencil, ink, or watercolor to serve as a guide in cutting the stencil, the number of colors used determining the number of stencils to be cut. The other procedure lay in combining separate stencils delineating lines, curves, and solid portions of the finished design. This finished design was predetermined only by its purpose, shape, and size; the actual form was the result of sometimes accidental, at other times inevitable, combinations of the elements.

TIES, breaks in the design, are essential to holding the stencil intact. They may either be filled in after the design has been printed; or, in multicolorwork, they may be covered by succeeding colors; or they may be planned as an integral part of the design. On the sketch, ties may be indicated "with white watercolor as warnings not to cut across."

7

TO CUT A STENCIL from a sketch, first tape the sketch to white cardboard; then, over the sketch, tape the film to be cut. Thus the mounted sketch and film can be turned freely as the cutting requires. For multicolor designs register marks must be added.

"The cut in the stencil film is made by pushing the knife away from you. Begin cutting in a central region of the design, working outward to the edges. Watch the drag of the point of the knife in the film. The first cut is made lightly, not much more than a scratch." When a piece of black paper is inserted between the sketch and film, it shows the scratches as white lines. To make them more visible, rub with French chalk. Then carefully recut these lines to penetrate the film. At times, if the scratch is merely deepened, bits of the portions to be removed can be *pushed* out with a knife.

Small stencils for monochrome printing can be used without any sort of MOUNTING, although extra care in handling is required. However, if a stencil device is to be used often, it is desirable to tape it to the bottom of a piece of cardboard in which an opening has been cut slightly larger than the design. The card serves as a convenient "handle," helps to hold in place the material to be stencilled, and makes it visible as it lies on the drawing table. A clean, transparent stencil can be very elusive!

Stencils for multicolor designs must be mounted on a printing frame and the stock fed to gauges. It is sometimes more convenient to print monochrome designs in this way.

If much stencilling is to be done, it may be preferable to make a wooden printing frame. A wooden rectangle similar to a picture frame is fastened with two hinges to a drawing board. Cardboard, the same size as the frame, is tacked to the bottom of the frame. After an opening slightly larger than the design is cut into the center of the cardboard, the stencil is taped to the underside, over the hole.

A simple but efficient printing device can be made from cardboard. An oblong corrugated board is bent once or twice along the shorter dimension about a third of the way from the end. The smaller area is glued to a piece of heavy cardboard with the in-

8

OVERLEAF: *Stencilling Equipment*

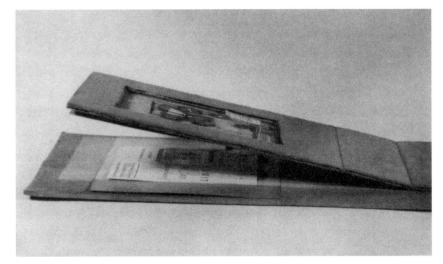

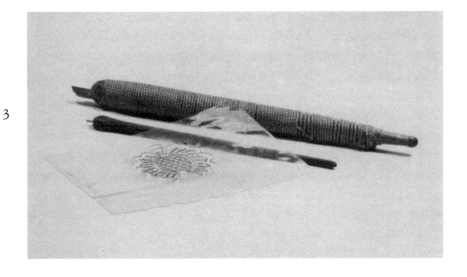

1

2

3

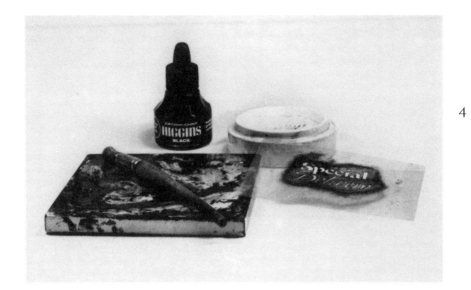

4

5

1. Wooden stencil printing frame, stencil in place, paper fed to gauges

2. Cardboard stencil printing frame

3. Stencil, needle-point drill, and handmade knife

4. Ink, china slab for drying ink, brush, water dish, and stencil after use

5. Stencil and two stencil brushes, the larger of which is used for pochoir

side of the fold upward. The opening for the stencil is cut into the wider area.

After a piece of stock has been positioned under the stencil to determine the location of the gauges, holes are cut into the cardboard of the hinged frame in order to allow the gauges to project.

"All the PIGMENTS are water-soluble. For black, Higgins' waterproof drawing ink; for transparent colors, tube watercolors; for opaque colors, 'Chinese white' tinted with tube watercolors, or poster colors of good quality. With a palette knife, pigments are spread out thin on glass plates and allowed to dry. To load the brush with just enough pigment (and not too much!) the tip of the handle of the brush is dipped into a cup of water, a drop or two snapped off onto the plate, and the pigment scrubbed up. The charge that the brush carries is a critical point in the stencil operation: too much ink or color on the brush will run under the the stencil; too dry a brush" will produce a poorly defined print, possibly damaging the paper as well by over-scrubbing.

TO WASH A STENCIL, moisten it under the faucet, place it flat on the sink, sprinkle with fine cleansing powder, rub with ball of finger, and rinse. Dry by placing it flat between paper towels.

9

DESIGNS FROM WOODBLOCK ELEMENTS

THE woodblock process was first described by Dwiggins under the pseudonym of Hermann Püterschein in *Transactions of the Society of Calligraphers*, Bulletin No. 2, 1925:

> *This method . . . is the by-product of an enthusiasm . . . for the stamped congeries of sprigs and flowers that embellish old Indian printed cottons. It suggested . . . that a like naive effect might come from cutting single leaves and buds on wood and stamping them in drawing-ink on paper. . . . They are cut on bits of maple-wood "on the plank" and are manipulated as rubber stamps are manipulated . . . . I am informed that the uncertainties incident to the process are maddening.*

11

Shown here are four designs made by this process, together with some of the stamps (slightly reduced)used in their production.

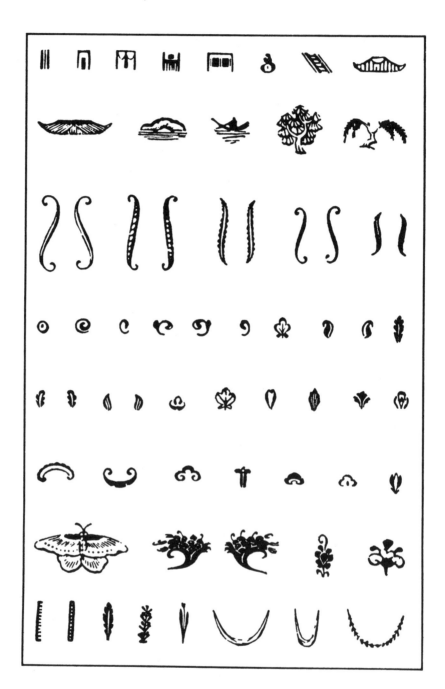

12

13

On decorating contemporary books with ornaments of contemporary inspiration, Paul Beaujon observes:

> . . . there assuredly are abstractions behind the cold perfections of mechanical design that can give rise to a system of ornament. Mr Dwiggins phrased it happily: "The thing is not to draw the working machine, but the curve that is in the mind of the engineer before he designs the thing." This can be done: Mr Dwiggins has done it. There is a moment, he argues, in which the mechanical designer confronts his problem and finds in his visual memory a vast equipment of potential movements, lines, curves, not of metal parts that he has seen but of the austere bodiless beauty of higher mathematics: gestures in space that imply duration, stress, movement. . . . Mr Dwiggins has made raids on this unconscious studio and brought out a number of approximations in line . . . . Once these are put on paper, the problem is how to arrange them. . . . growth-form is something natural and inevitable to us; and Mr Dwiggins is justified for this reason in combining his engineering abstractions into that most familiar mechanical pattern, the structure of the plant. Hence the "vignette" in his hands is somehow still the growing vine, but the result is spoken as it were in the present-day vernacular instead of by the false naturalism of a past day.

This is quoted from "On Decorative Printing in America," *The Fleuron,* No. 6, London, 1928, which includes a discussion of Dwiggins' ornaments from the point of view of design.

15

16

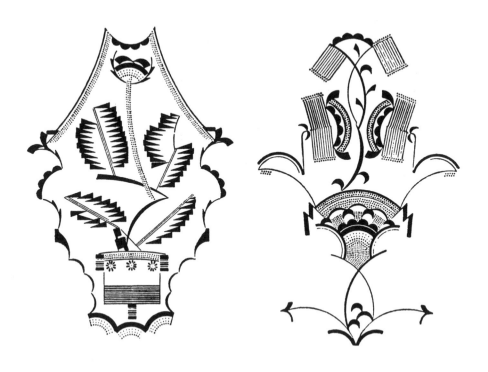

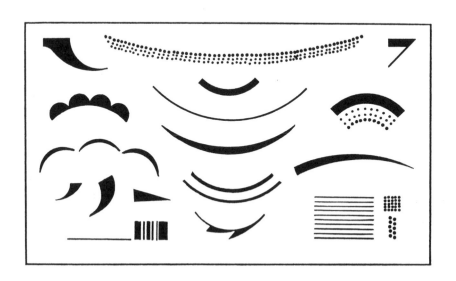

17

18

19

20

21

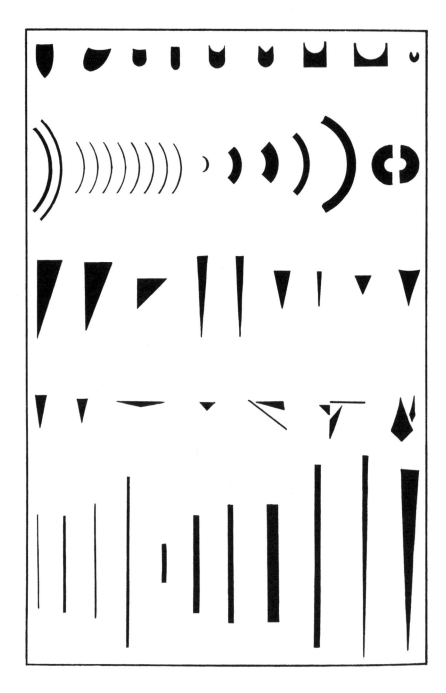

22

23

24

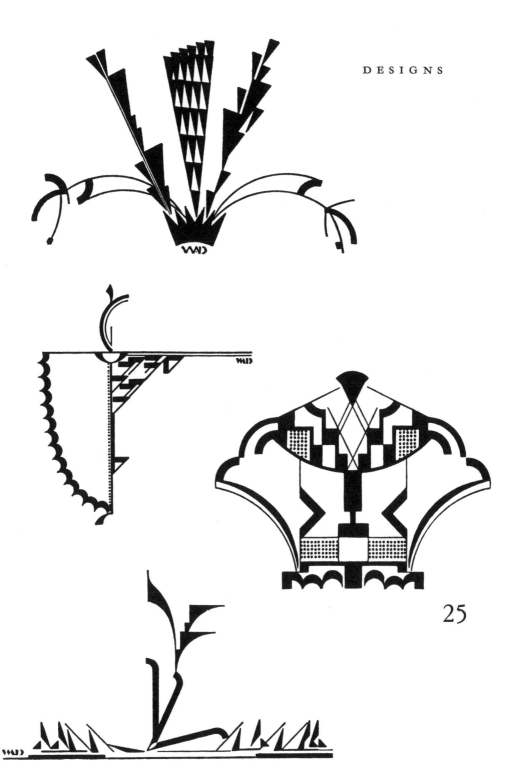

25

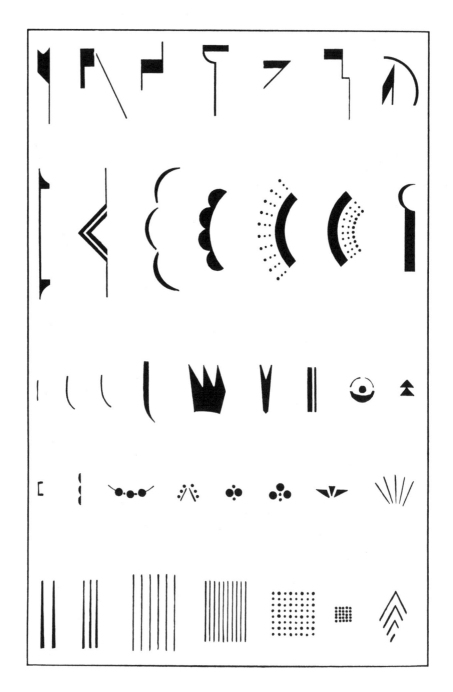

26

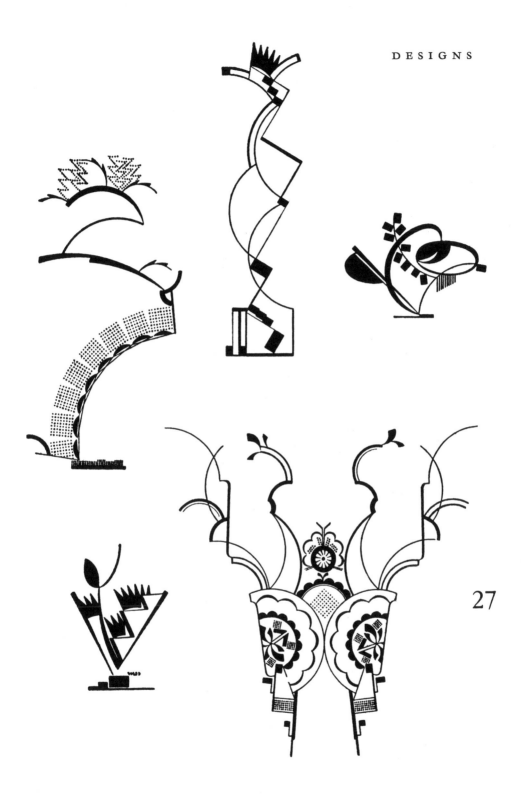

27

28

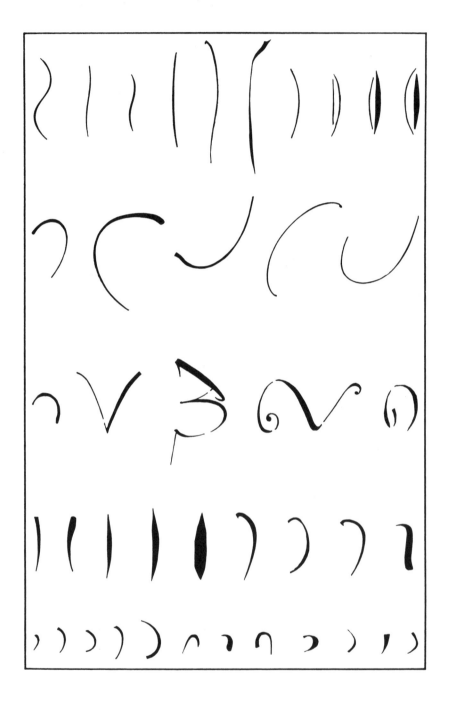

29

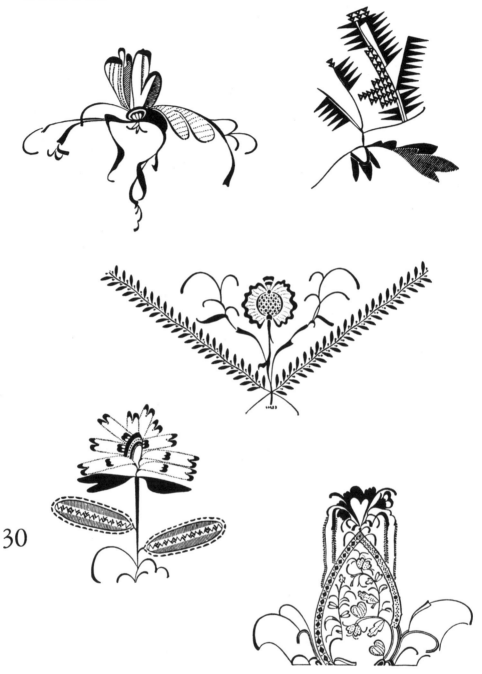

30

31

32

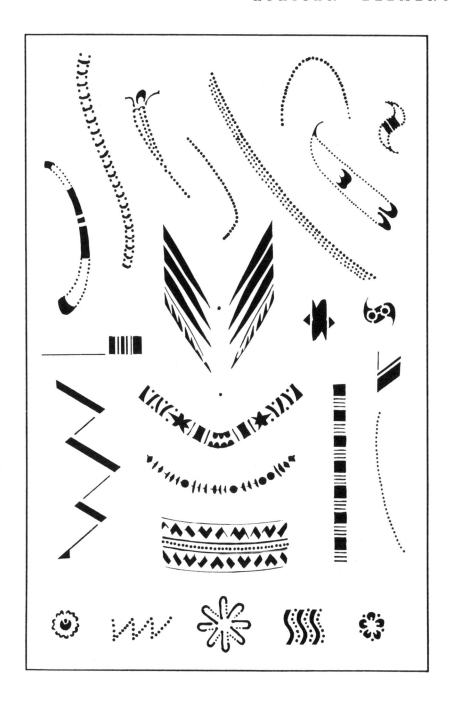

33

from

the standpoint

of calligraphy

he is painting a

twelve-league

canvas

with brushes

of comet's hair

A certain activity of Mr. Dwiggins was thus described by Paul M. Hollister. It is pertinent to quote at greater length:

34

A year or two back, if you had gone into Mr. Dwiggins' studio, you would have found him on a high stool, filing drill-edges on a needle-point. When the needle-point had become a drill, it was mounted in a stick. Mr. Dwiggins then poised the needle over a sheet of celluloid, and twirling the drill between his palms as a West Indian bar-tender twirls a swizzle stick, he would drill a pinhole in the celluloid. He drilled rows of pinholes, and then more rows, and when he got tired, he drilled pinholes. Clients telephoned for their drawings. Mr. Dwiggins replied sweetly that they were almost ready—and went back to drilling fly-specks.

When he had enough fly-specks drilled, he simply used the celluloid for a stencil, and brushed ink on paper through the pinholes. The result, naturally, was a pattern of fly-specks.

What of it, says you. Just this: neither pen nor brush in human hands can yield on paper precisely the quality of fly-specks drilled through a stencil with a needle. The pen picks the paper, or flows the ink from a point; the brush-hairs buckle uncertainly. But there is a quality to the stencilled dots which pleases the eye . . . .

I mention it . . . because it is a fundamental symptom of the quality of his work. Mr. Dwiggins sits on a high stool and makes dots—but, from the standpoint of calligraphy, Mr. Dwiggins is painting a twelve-league canvas with brushes of comet's hair. He is one of the few men who can respect the simplicity of a fly-speck, appreciate its possibilities in design, and . . . . treat even dots in a new way.

—"Direct Advertising," Vol. xiv, No. 3, 1928

35

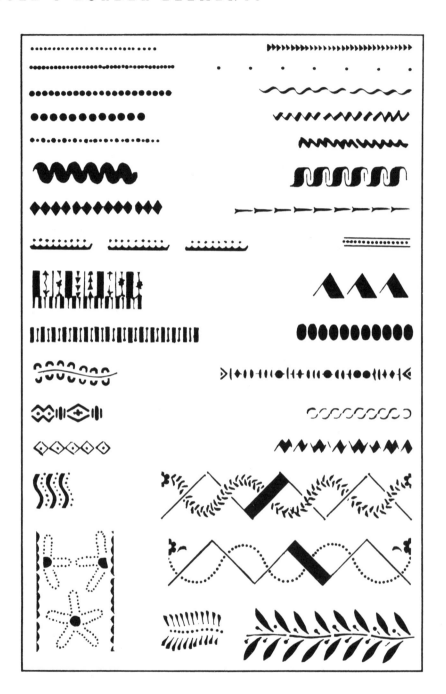

36

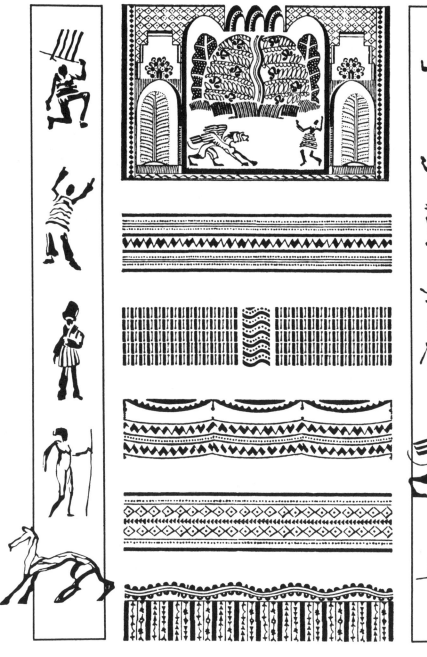

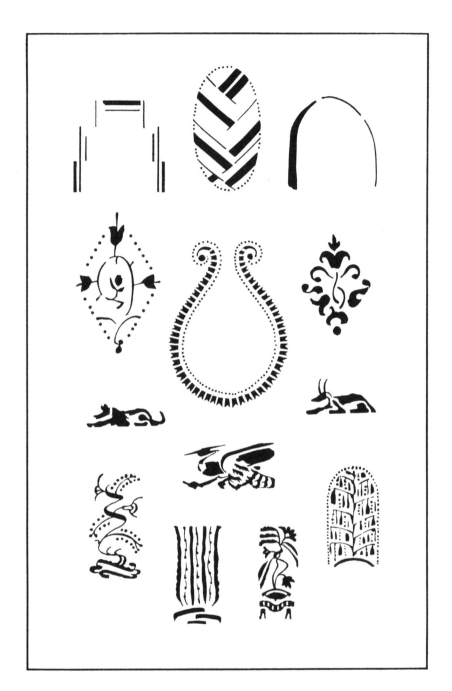

38

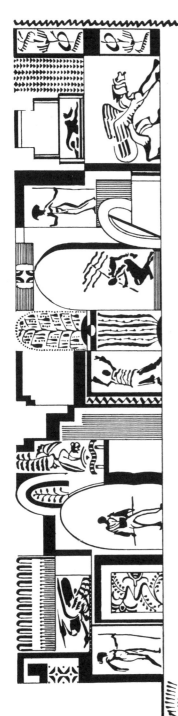

*the
vertical
side
and
bottom
borders
on
THE
TIME
MACHINE
double-
spread
title
page*

39

40

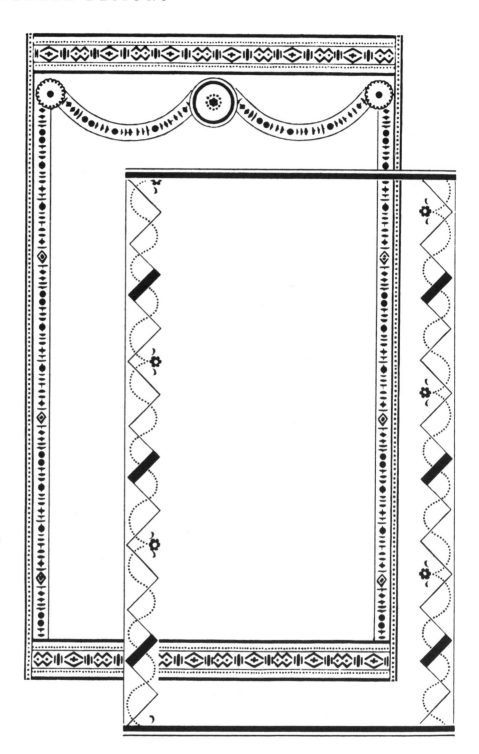

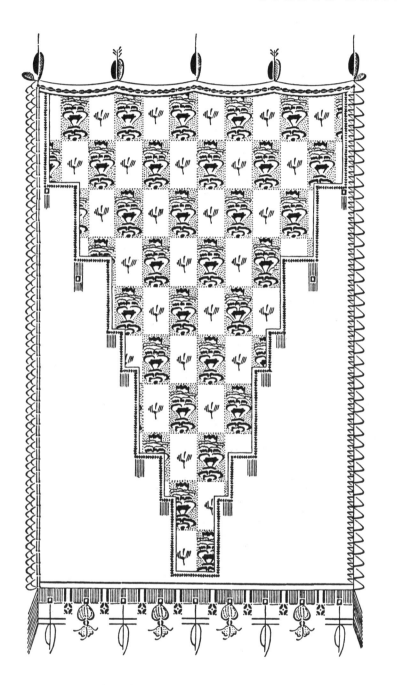

41

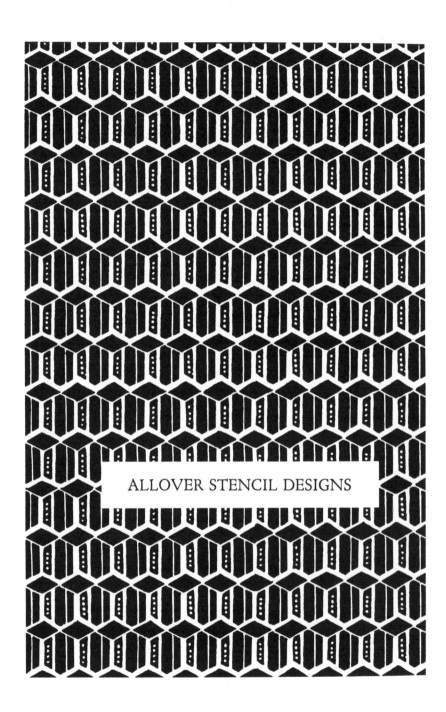

ALLOVER STENCIL DESIGNS

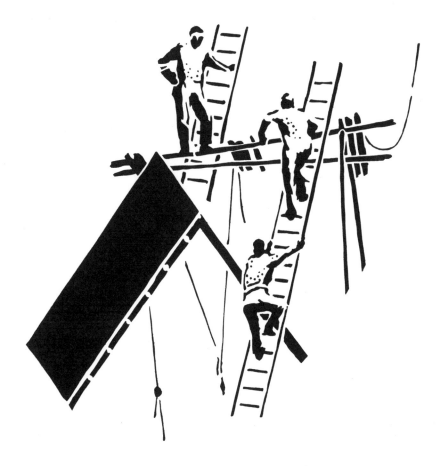

Dwiggins used his allover stencil designs chiefly for creating
patterned papers for bookbindings. A small section of the design
was printed, or *discharged,* directly through the stencil—small
elements onto graph paper; larger, onto 3-ply Strathmore plate,
which he customarily used for stencilling. Ties were filled in if
necessary. From this small section the lithographer then built up
a negative to the desired size. Occasionally these allover designs
were stencilled in their entirety.

43

44

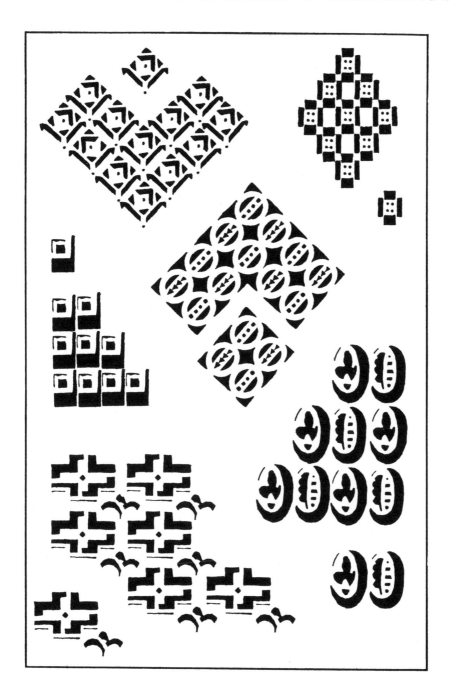

45

ELEMENTS

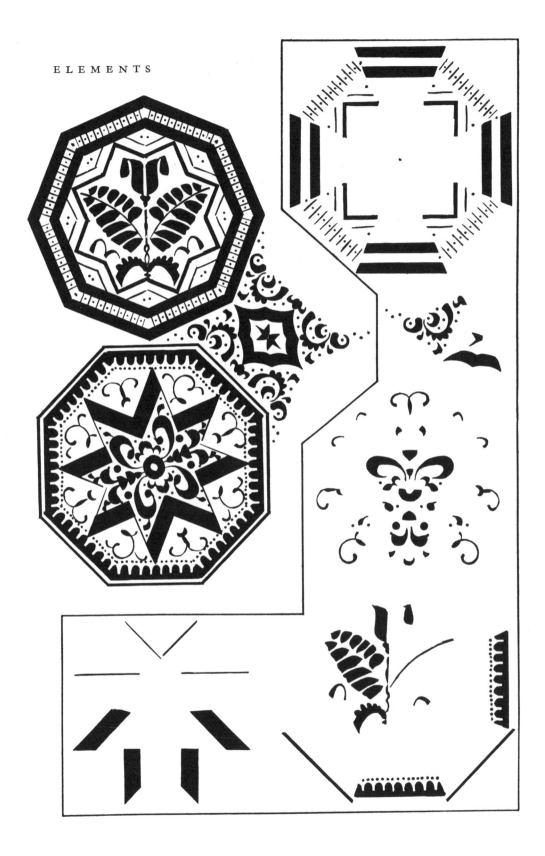

47

48

UNITS FOR ALLOVER PATTERNS

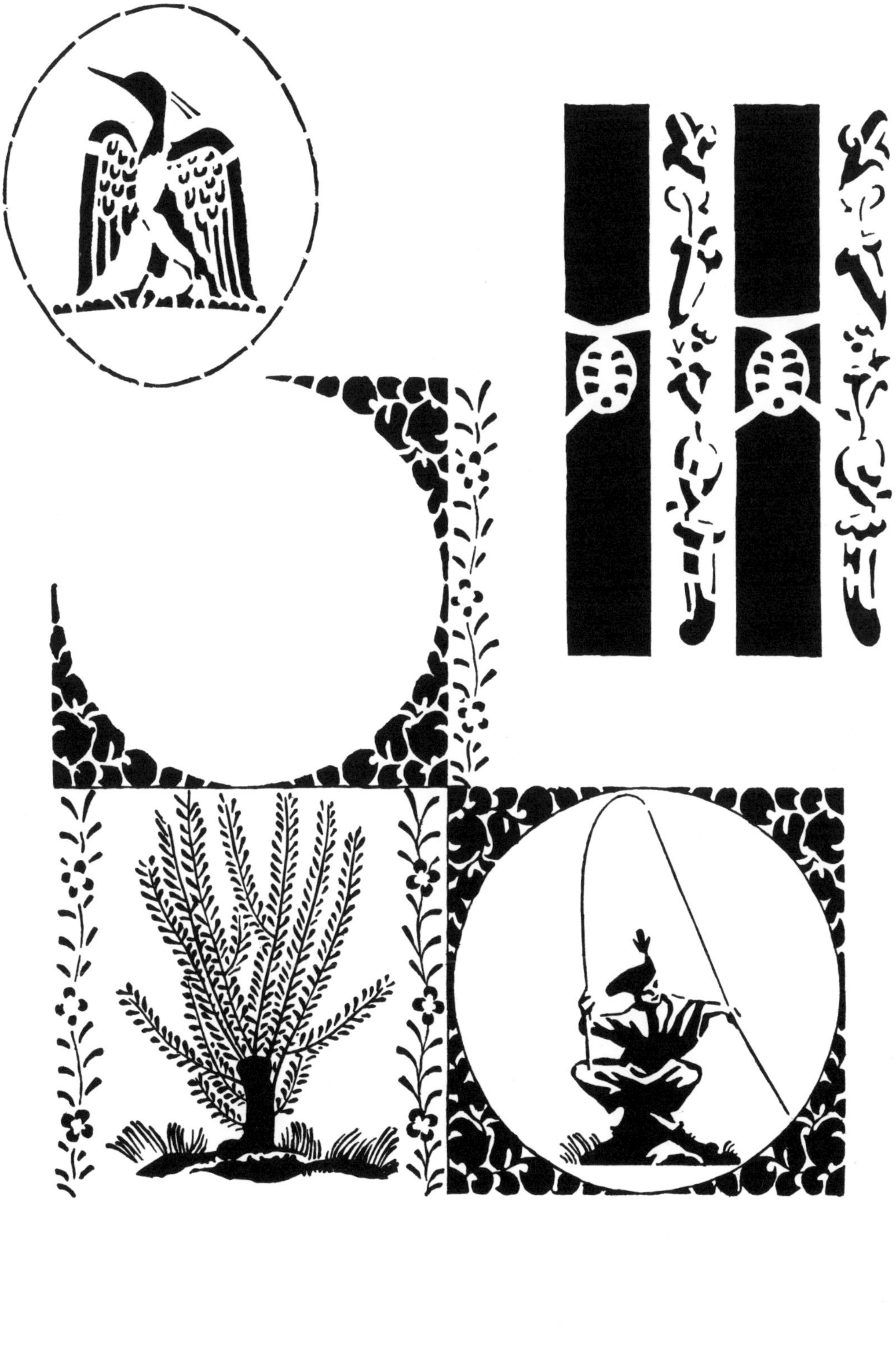

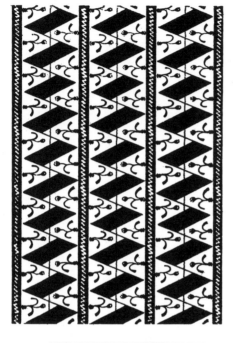

50

51

STENCILLED WORDS AND MUSIC

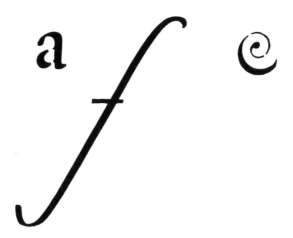

REGISTERED
Receipt requested

THE stencil was as well a practical tool in Dwiggins' hands. For daily usage there were various postal instructions to be printed on Manila envelopes. These stencils, along with his calligraphic return-address labels, complemented the recipient's address in-inscribed in Dwiggins' highly individual handwriting—all three contriving to make a collector's item of a plain Manila envelope.

Stencilled letters were used for words and phrases, as well as for captions for some of his stencil "prints," print being used in the sense of a multicolor woodcut.

The stencil facilitated his work in type design. As an aid in arriving at the first-run experimental letters for Falcon, he cut stencils for "a long and a short stem, the *n* arch, and a loop—twice the size of 12 point—pretty small!—and constructed letters from these elements by stencilling." Later, as the face evolved, he cut cardboard templets (equivalent to large stencils) for the same stems and curves in the pattern drawing size.

There are several alphabets of various sizes and designs, most of them incomplete, which Dwiggins presumably used in circumstances where the quality of a stencilled letter was preferable to a drawn letter. One of the most useful of these alphabets, Imperial, is in approximately 120 point size.

Dwiggins' long-standing interest in the design of musical notation took concrete form in experimental work for a music publisher. And so it is not surprising to find among his stencils notes and clefs, sharps and flats, and other musical signs.

TO   BE

Registered

❋ receipt requested

54

Special
Delivery

Special delivery

First class

55

Three Designs for
SINBAD
published by
The Society of Calligraphers
1925

Topographical Overtones

56

Copies, №   P A   WAD

Published by
M. Firuski, Dunster House, Cambridge, Mass.

Graphic response to the stimulus **CONGO**

Fifty Copies, №   WAD   57

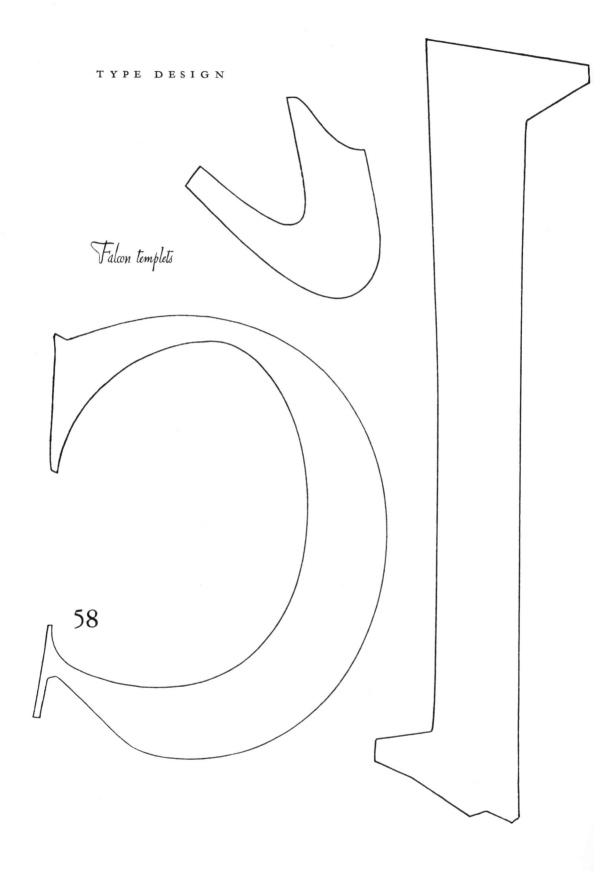

Falcon templets

58

Falcon stencils I ı ˥ ɔ c

nihil diminuendum

59

A B C D E I K M
N O P R S T U V

a b c d e f g h i j k l m
n o p q r s t u v w x y z

a c c c f g i i l o i r s t v y

60

e e g n i l
i o o r s t

A B C D E F G H I J K
L M N O P R S T U V Y

a ɔ c l g ɔ ı o t

a b c d e f g
h i l m n p q
r s t u v

61

afgil
smorst

acefgt

62

ayʃʙʋȝʏʃʀstʋ

ABCIDEIKM

NOPRSUVY

einrst

ABCDEIMNORSTU

FMCSTW   63

AIDEMNOPRSTU

ABCD
EIMN
OPRS
TUV

64

CAPS
lower case

alcefgik
ojc˘stuy

65

66

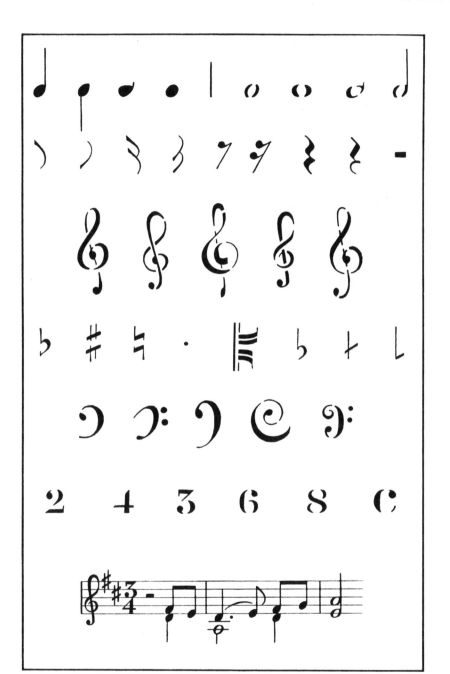

67

POCHOIR ILLUSTRATION

Pochoir, French for "stencil," is a method of reproducing paintings by means of copper or zinc stencil plates. In Paris this technique has been highly perfected, often employing thirty or more stencils to achieve the precise effect of the original art.

Dwiggins used pochoir, with celluloid stencils, for the illustrations in several books, but in so doing he regarded it as a creative medium, not just a reproductive process. ". . . he wanted to get the sharp, positive, simplified effect that results naturally from the stencil method." *The Treasure in the Forest* is so produced. Each picture is an original in the sense that a woodcut print is an original work of art.

For *The Time Machine* and *One More Spring* he "worked out his color-scheme with the stencil plates and then made discharges from the stencils in black drawing ink to serve as 'copy' for the process engraver." Therefore these books contain reproductions of his original drawings.

In addition to the book illustrations, he made what might be called pochoir "prints:" episodes from *Sinbad,* and, as he designated them, graphic responses to verbal stimuli, "La Paz" and "Congo," the latter utilizing twenty-two colors.

The stencils for three unpublished illustrations indicate that he had considered but decided against the use of pochoir for two other books.

69

NOTES AND SOURCES

SOME of the designs shown in this book have been reproduced from copy made directly from stencils; others have been photographed from the books and pamphlets in which they appeared, resulting, at times, in ragged or imperfect reproductions.

As it was not feasible to reproduce the designs in the original colors, those that were not printed in black are so indicated.

The books, pamphlets, or other sources of the ornaments and illustrations are listed below, followed by the page numbers on which the designs appear in this book. At times more than one design comes from the same source. The designs from the same source are listed together. Since these designs do not necessarily appear on consecutive pages, the sources are listed sequentially according to the first page in this book on which a design from it appears.

71

Abbreviations: *t*, top; *c*, center; *b*, bottom; *l*, left; *r*, right

TRANSACTIONS OF THE SOCIETY OF CALLIGRAPHERS, Bulletin No. 2, February 3, 1952, Boston

Page 3 *headpiece*

WARREN'S STANDARD PRINTING PAPERS, S. D. Warren Co., Boston

Thirty paper samples, each with a different design, each design printed in one of fifteen colors used in the portfolio

4   sample 13 *l*
24   samples 27 *t*, 19 *lc*, 24 *rc*, 28 *b*
25   samples 1 *t*, 9 *lc*, 15 *rc*, 12 *b*
30   sample 7 *c*
31   sample 17 *c*

TALES, by Edgar Allan Poe. The Lakeside Press, Chicago, 1930

4   "Morella" *r*
16   "The Gold Bug" *t r*, title page *c*
30   "The Purloined Letter" *t l*, "Berenice" *t r*, "Ligea" *b l*,
      "The Assignation" *b r*
31   "Hop-Frog" *t*, "The Fall of the House of Usher" *b*

PARAPHS, by W. A. Dwiggins. Alfred A. Knopf, New York, 1928

Text ornaments in orange vermilion

5   "The Last Mobilization" *l*
17   "Traverse" *l*, "Comment on . . . Dynamic Symmetry" *r*
27   "Inconclusive Incident of Mr. Wang" *t l*
      "The Tympanum Hypothesis" *t c*
40   title page *r*
50   binding design; mars violet *t l*

72

Invitation to the AIGA Dwiggins Exhibition. New York, 1937

5   ornament; gray green   *r*

S. D. Warren Company Booklets

6   "Anybody's Prospect or, 'I'll Stick to the Old Bus;'" magenta
14   Olde Style - Antique Laid, ornament detail; orange

Unpublished Designs

9   ornament, *tailpiece*
34   ornament, *headpiece*

Page 35   ornament, *tailpiece*

47   binding design for Poe's *Tales*

MODERN COLOR, by Carl Gordon Cutler and Stephen C. Pepper. Harvard University Press, Cambridge, Mass., 1923

11   Chapter 5

13   Chapter 4 *t,* Chapter 9 *b*

Announcements for Charles Hovey Pepper Exhibitions. Doll & Richards, Boston, 1923, 1929

13   "North Country;" Prussian blue *c*

21   "North Country;" mars violet *t*

THE FLOWERS OF FRIENDSHIP, edited by Donald Gallup. Alfred A. Knopf, New York, 1953

20   binding design; gold stamping *t*

THE ARCHITECT AND THE INDUSTRIAL ARTS. Metropolitan Museum of Art, New York, 1929

20   title page *b*

21   tailpiece *c*

CHELKASH AND OTHER STORIES, by Maxim Gorky. Borzoi Pocket Books, Alfred A. Knopf, New York, 1923

21   detail of jacket design; chrome green *b*

THE HOUND AND HORN, Vol. III, No. I. Portland, Maine, 1929

27   tailpiece *t r,* tailpiece *b l*

ATTITUDES, July, 1927. The Caxton Co., Cleveland

27   title page; yellow ochre *b r*

THE TREASURE IN THE FOREST, by H. G. Wells. Press of the Woolly Whale, New York, 1936

37   headpiece for opening page *t,* five headbands for text pages *c* to *b*

THE TIME MACHINE, by H. G. Wells. Random House, New York, 1931

39   title page borders: these were printed vertically, as here, along the outer margins of the double-spread; raw umber

41   half title design

Poster for the Society of Printers Meeting. Boston, December 9, 1927

40   border design for poster, fifteen inches long *l*

Patterned Paper for Paper-over-Boards Binding for Knopf

42   one of several designs in various brilliant colors

74

The text of STENCILLED ORNAMENT & ILLUSTRATION
has been hand set in Winchester Roman,
an experimental Linotype face designed by Dwiggins.
It was cut in the 12 point pilot size only, lacking italic caps.
Garamond has been employed for the folios and
supplementary material

Reproduced from the original Püterschein-Hingham edition
designed and printed by Dorothy Abbe

AFTERWORD

——

BRUCE KENNETT

When I think of the expression "a labor of love," I cannot imagine better examples than this book and its very different sister volume, Dorothy Abbe's *The Dwiggins Marionettes*.[1] By the time Dorothy produced the first edition of her stencil book, twenty-five years had passed since she and W. A. Dwiggins first began to imagine it, yet her devotion to his artistic legacy remained unchecked. Now, after an additional thirty-five years, Rob Shaeffer and Princeton Architectural Press have shown another kind of devotion, making possible a new and expanded edition, faithful to Dorothy's original but with the significant addition of color reproductions of Dwiggins's stencils.

Dorothy worked closely with Dwiggins during the last ten years of his life, her presence and industry enabling him to produce far more work than if he had attempted to do it on his own; in the twilight of his career, already in his seventies, he was challenged by palsy and cataracts, in addition to coping with diminishing reserves of energy after more than thirty years battling severe diabetes. For those of us who knew Dorothy and spent time in conversation with her, the real wonder was that she had spent *only* ten years with him—so thoroughly did she know the breadth and depth of his output. After Dwiggins's death, Dorothy devoted herself to the preservation of his memory: she gave talks and wrote articles about his work, organized his papers, and established the Dwiggins Collection at the Boston Public Library. Dorothy's books on Dwiggins also reveal her own superior skills and craft as typographer and photographer.

Dorothy was born in 1909 in Dorchester, Massachusetts, a town located just south of Boston and less than ten miles from Hingham, where Dwiggins lived from 1904 until his death in 1956. After graduating with a

77

78

Early on, Dwiggins used stencils to add flat areas of color to his black-and-white designs. **1.** Holiday card published by Alfred Bartlett, 1908. **2.** Calligraphic holiday card for sale by Dwiggins, 1908 (margins reduced). **3.** Proof for bookplate, 1916. [ALL: BPL]

major in Latin and a minor in printing arts from Randolph-Macon Woman's College, Dorothy taught high school Latin for a few years, worked in a bindery, and eventually set herself up as a designer of books. She first met Dwiggins in 1933. Their first project together was a book for Dwiggins's old friend John Phillips, *The Bomb That Wouldn't Go Off*, for which Dwiggins made the illustrations and Dorothy the typographic design.[2] By 1946, their lives had grown increasingly connected, as she became not only Dwiggins's assistant but also a live-in helpmeet to both Dwiggins and his wife Mabel, who was suffering from asthma and early dementia. After Dwiggins died on Christmas Day 1956, Dorothy continued to care for Mabel until she died twelve years later. Dorothy then made her home in Dwiggins's studio on Irving Street in Hingham, which he designed and had built in 1937. Those of us who knew Dorothy cherish memories of our visits with her in this special place. For anyone interested in learning more about Dorothy, I enthusiastically recommend Anne Bromer's *Strings Attached: Dorothy Abbe, Her Work and WAD*.[3]

In producing her first handmade edition of the stencil book in 1979, Dorothy's goal was to present the material that she and Dwiggins had always planned to publish. She undertook this on her own, setting the text by hand in Dwiggins's Winchester type, and then printing it herself (again by hand) on her small press. Therefore, her entire production was minimalist, and, by definition, printed only in black ink. A 1980 reprint published by the Boston Public Library retained Dorothy's black-only design. If a reader were to come upon either earlier edition of Dorothy's book, he or she would have no idea of how joyful Dwiggins was in his use of color. He once wrote, "I like Far East color combinations; a chutney sauce effect with lots of pepper and mustard and spices, off harmonies that make you sit up. I think the Chinese were the greatest color manipulators, and after them the Persians of the miniatures." And indeed, Dwiggins's love of stencils also has ties to China: the earliest examples known to us are found in the western part of that country, in the Caves of the Thousand Buddhas near Dunhuang.

79

4. By the mid 1910s Dwiggins had begun to use wooden stamps to produce complete illustrations. The image shown here accompanied a short story by Lord Dunsany in *The Fabulist 2* (1916), an occasional journal published in small editions by Dwiggins and his cousin Laurance Siegfried between 1915 and 1921. [BPL]

As we look back at his work from the perspective our own time, Dwiggins remains unique in his use of what he called "faded bright colors" and in the sheer inventiveness of his stencil designs. Even the most abstract examples reveal an organic, plant-like quality of form. His color selections are unconventional, sometimes to the point of being jarring, and seem as if they shouldn't work, and yet most of the time they carry the day with clarity and playfulness. In making his stencils primarily to decorate and illustrate books, Dwiggins offered flashes of color that revealed his passionate interests in theatrical lighting and set design.

Beyond notions of color, Dwiggins learned directly from his own experiences that illustrations made to accompany text need to be rendered with qualities of line and shape that are in sympathy with the basic attributes of type characters themselves. This was especially true of the era in which he lived, during which most text was reproduced by letterpress. At that time watercolor washes or pencil drawings had to be reproduced by halftone on shiny coated paper, and thus were not a successful marriage with text, whereas woodcuts or stencil images were an ideal match. They could be printed on uncoated paper with great success, and often at the same time as the type.

Over the last thirty years of his life, Dwiggins developed nearly a dozen typefaces for Linotype, of which five reached full production and release. Electra and Caledonia remain the best-known of these designs; they became huge commercial successes and enabled the Linotype machine to become a viable choice for book composition. As his skills as a type designer deepened, Dwiggins made frequent use of stencils as he developed the shapes of strokes and characters (see pages 58–59). In 1938 Linotype issued Dwiggins's Caravan Ornaments, a series of decorative units that could stand alone or be made into countless combinations of lines or patterns. (For many years the working title for this series was "Chinese Spinach"!) Dwiggins's goal here was to build ornaments from the same lines and curves that are found in printing types, so that they could achieve the greatest harmony with printed text. As Dwiggins described them, it was

81

"as though particular groups of words had been told off for special ornamental duty, and had blossomed at command" into these decorative units.
Although Dorothy did not include any examples of Caravan in the book, an
indication of Dwiggins's approach may be seen on page 36, and the Caravan
units themselves are easily located via online image search.

   As an artist, Dwiggins commanded a wide range of techniques and had
a solid grounding in art history. He could complete one project in an angu-
lar art deco style, the next in chinoiseries reminiscent of Pillement, a third in
the spirit of eighteenth-century French engravings, another in the style of
Daumier. Dwiggins's stencil techniques provided him with the perfect tools
to serve his broad range of expression, from the quasi-realistic to full-on

82

5. "The Incandescent Cheese," watercolor, 1918. To avoid the tedium of painting the
wallpaper pattern by hand, Dwiggins used wooden stamps. (See pages 12–13).
He submitted this and another painting to the Boston Watercolor Society (BWS) as a part
of his application for membership; of all the entries in the 1918 BWS annual exhibition,
it's no surprise that this painting received the most attention. [BPL]

abstract. As his work evolved from wooden stamps to individual shapes cut from celluloid, he discovered that he could combine a few elements in myriad ways, while always providing a bit of imperfection and variety in the design, never something purely mechanical. In this, his humanity—and his Puckish sense of humor!—always shone through. Posing as author Hermann Püterschein (Dwiggins's alter ego and *nom de plume*) in the 1939 booklet introducing the Caledonia typeface, he wrote: "[Dwiggins] creates an illusion of machines. But his machines are a masquerade. There are men inside them."

As you explore Dorothy's survey of Dwiggins's work, you will see many examples of personal projects as well as those produced on commission for clients. Running through all of them is the exuberant spirit of this creative and prolific genius, whose work in decorative art deserves much more recognition than it has had over the past fifty years. A delightful example of Dwiggins's *joie de vivre* may be found in this account of his own working methods:

> You take a cork out of the top of your head, and you drop in a word like La Paz, or Congo, or Sinbad. One word at a time. If it's the name of a place it need not be a place you know. If it's not the name of a place, but just a word, you need not know it so fine as to split hairs. Just put the word in. Then put in a couple of cocktails and some black coffee, and put the cork back in tight, and jump up and down for two or three days and then the word will come out of your fingers onto the paper. Then you give the result—picture or pattern, or whatever it is—a high-sounding caption like *Graphic Response to Verbal Stimulus: La Paz*. That's all there is to it. It doesn't mean a thing but it's a lot of fun.[4]

83

1 Dorothy Abbe, *The Dwiggins Marionettes* (New York: Abrams, 1970).

2 John Phillips, *The Bomb That Wouldn't Go Off* (Boston: Bruce Humphries, Inc., 1941).

3 Anne Bromer, *Strings Attached: Dorothy Abbe, Her Work and WAD* (Boston: Boston Public Library / Society of Printers, 2001).

4 From Paul Hollister's preface in *The Work of W. A. Dwiggins Shown by the American Institute of Graphic Arts at the Gallery of the Architectural League* (New York: American Institute of Graphic Arts, 1937).

THE

Time Machine

H. G. WELLS

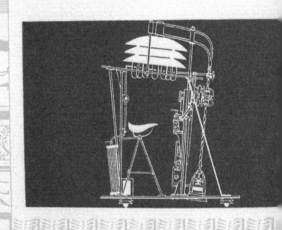

AN INVENTION

*With a preface by the Author
written for this edition; and
designs by* W. A. Dwiggins

Random House *New York*

Topographical Overtones   CONGO   25 Copies, № 7   W. A. Dwiggins

86

PREVIOUS PAGES: **6.** Title-page spread from *The Time Machine*. Dwiggins created
the decorations with myriad stencil elements cut in celluloid, using black ink to generate
artwork that would be printed in color by letterpress (see pages 37 and 39).
All text was lettered by hand. [LA] **7.** Study for a multiple-color stencil print, 1926.
Between 1924 and 1926 Dwiggins produced a portfolio of large prints which illustrated
some scenes from Sinbad's voyages, but he also expanded his explorations
to include other faraway places. [BPL]

STREETS

IN

THE MOON

BY
ARCHIBALD MacLEISH
Author of "The Happy Marriage" and
"The Pot of Earth"

BOSTON AND NEW YORK
HOUGHTON MIFFLIN COMPANY
The Riverside Press Cambridge
1926

87

8. Title page from *Streets in the Moon* by Archibald MacLeish (Boston:
Houghton Mifflin, 1926.) Dwiggins built this artwork using wooden stamps and India ink,
combined with refinements added in pen. This is probably Dwiggins's highest
level of achievement with the wooden-stamp technique. The problem with the stamps
was that it was difficult to see exactly where the impression was being made
in relation to other elements already placed on the page, and the stamps could not
be flopped to create a reverse image. [BKC]

88

9. In the late 1920s Dwiggins pitched several of his friends in publishing circles,
hoping that they would accept a volume of his "Athalinthia" fantasy stories. This artwork
was prepared as a covering for the front and back boards of the hardcover book
he proposed to design. His notation (written in later years) reads, "I made this stencil one
time as a possible side-paper pattern—but paper sides & cloth back seem
to be out in the trade." [BPL]

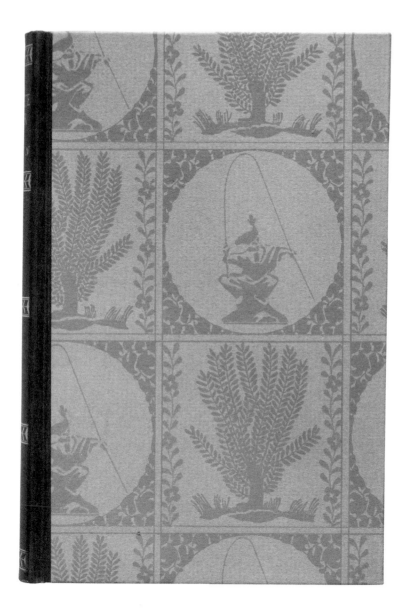

89

10. Paper-covered front and back boards for a 1928 edition of Isaac Walton's
The Complete Angler, designed by Dwiggins for Boston bookseller Charles Goodspeed.
Dwiggins once described this cover design as "Dutch tiles with a Persian twist."
(See page 50.) [BKC]

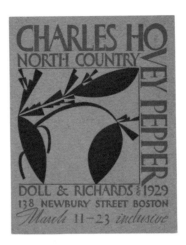

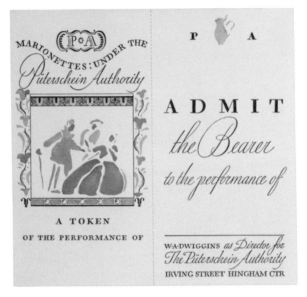

90

ABOVE: **11.** Exhibition notice for artist Charles Hovey Pepper, a long-time client, 1929.
Stencil ornament and hand lettering, printed letterpress in two colors. (See page 21). [BPL]
BELOW: **12.** Souvenir ticket for a performance at Dwiggins's private marionette
theater, ca. 1933. Hand lettering and frame decoration printed by letterpress in black, colors
then added by hand via stencil (see photo 2 of stencilling equipment, following
page 8). Dwiggins was a skilled maker and engineer of marionettes, and was well-known
in the puppet world; he also designed stage sets, lighting, and costumes. [BKC]

13–15. Sample chapter opening (*above*) and two other examples of headpiece artwork for a limited edition of Robert Nathan's *One More Spring*, designed and illustrated by Dwiggins and published by The Overbrook Press in 1935 (see page 68 for another example, as well as the jacket of this 2015 edition.) The artwork was made by hand in black, then printed in multiple colors via letterpress from photo engravings. [BKC]
FOLLOWING PAGES: **16.** A spread from *The War Against Waak*, one of Dwiggins's fantasy stories published by Püterschein-Hingham in 1948. Dwiggins and Abbe established the P-H imprint to publish small and colorful projects that were of particular interest to them. In this case the illustrations in each book were colored by hand via stencil. [BKC]

91

Zond Sarfa

92

THE BATTLE

OPPOSITE ZOND

PÜTERSCHEIN-HINGHAM *wishes to thank its customers for their encouraging support.*

This support enables PH *to exist: it supplies the wherewithal to buy paper, ink, photo-engraving; packaging and posting.* PH *takes no account of the cost of design and drawing, typesetting and presswork.*

We like to experiment with designing and printing books . . . your support makes it possible for us to do so.

So we thank you.

DA & WAD

1952

94

17. Notice from the proprietors of Püterschein-Hingham, 1952. Dwiggins loved to make designs that combined natural and abstract details, and which often appeared to float in three-dimensional space. [BPL]

18. Manila envelope with samples of Dwiggins's hand-stencilled instructions to the Post Office (*center*) and printed return-address stickers and bookplates (*corners*). [BPL]

PUBLISHED BY:
Princeton Architectural Press
37 East 7th Street, New York, New York 10003
www.papress.com

ISBN 978-1-61689-375-0

AFTERWORD ILLUSTRATIONS:
The following are reproduced with the kind permission of their respective publishers:
Item 6, Illustrations by W. A. Dwiggins © 1931 W. A. Dwiggins. Copyright renewed 1959
by Random House; from *The Time Machine: An Invention* by H. G. Wells. Used by permission of
Random House, an imprint and division of Random House LLC. All rights reserved.;
Item 8, Title page from *Streets in the Moon* by Archibald MacLeish (Boston: Houghton Mifflin, 1926);
Items 13–15 and dust jacket, The Overbrook Foundation.

PHOTOGRAPHY:
Items 1, 2, 5, 8, 10, and 12–15 by Bruce Kennett; Items 3, 4, 6, 7, 9, 11,
and 16–18 by Letterform Archive.

ITEM CREDITS:
Dwiggins Collection, Boston Public Library [BPL]; Bruce Kennett Collection [BKC];
Letterform Archive [LA]. Deep thanks to Susan Glover, Sean Casey,
Kim Reynolds, and Roberta Zonghi of the Boston Public Library for their patient and
constant help over many years; and to Rob Saunders and Sylvia Davatz
—Bruce Kennett

EDITOR: Rob Shaeffer · DESIGN COORDINATION: Paul Wagner
Afterword typeset in Metro Nova Pro, designed by Toshi Omagari and William Addison Dwiggins

Special thanks to: Nicola Bednarek Brower, Janet Behning, Erin Cain,
Carina Cha, Tom Cho, Barbara Darko, Benjamin English, Jan Hartman, Jan Haux,
Lia Hunt, Mia Johnson, Valerie Kamen, Stephanie Leke, Diane Levinson,
Jennifer Lippert, Jaime Nelson, Sara Stemen, Marielle Suba, Joseph Weston,
and Janet Wong of Princeton Architectural Press
—Kevin C. Lippert, publisher

Cataloging-in-Publication data
available from the Publisher upon request.